YOUR KNOWLEDGE HAS VALUE

- We will publish your bachelor's and master's thesis, essays and papers

- Your own eBook and book - sold worldwide in all relevant shops

- Earn money with each sale

Upload your text at www.GRIN.com and publish for free

Tehseen Ahmad

Implementation of Variable Frequency Drives (VFD) on Boiler Feed Water Pumps for Drum Level Control

GRIN Publishing

Imprint:

Copyright © 2015 GRIN Verlag GmbH
Print and binding: Books on Demand GmbH, Norderstedt Germany
ISBN: 978-3-656-97151-1

This book at GRIN:

http://www.grin.com/en/e-book/298967/implementation-of-variable-frequency-
drives-vfd-on-boiler-feed-water

GRIN - Your knowledge has value

Since its foundation in 1998, GRIN has specialized in publishing academic texts by students, college teachers and other academics as e-book and printed book. The website www.grin.com is an ideal platform for presenting term papers, final papers, scientific essays, dissertations and specialist books.

Visit us on the internet:

http://www.grin.com/

http://www.facebook.com/grincom

http://www.twitter.com/grin_com

VFD Implementation on Boiler Feed Water Pumps for drum level control

The project was to convert the control of three boiler feed water pumps to Variable Voltage Variable Frequency (VVVF) drives having capacity of 750 m^3/H @ 220 Kg/Cm^2 pressure and power rating of 6200 KW each.

The main focus of this report is the design and development of the protection system, sequence of operation, bypass system, speed control system, drum level control and graphic interface. It also include PID controller tuning for VVVF drive smooth control.

Tehseen Ahmad CEng MInstMC CAP

I&C Engineer

This Report is dedicated to my parents & family

Executive Summary:

This is a technical report on the implementation of variable frequency drives (VFDs) for the boiler feed water pumps. It includes the planning, execution, calculations, simulation, testing and commissioning of the VVVF (variable voltage & variable frequency) drives.

The Lalpir Power Limited is an HFO fired thermal power plant having gross capacity of 365MW. It is situated Near Mahmud Kot, Tehsil Kot Addu, District Muzaffargarh, in the Province of Punjab, Pakistan.

The power plant is electrically connected with Water and Power Development Authority (WAPDA) system at 220 KV grid. It has one Generator Step-up Transformer (24 KV to 220 KV), one start-up transformer (220 KV to 11 KV), one auxiliary transformer (24 KV to 11 KV).

The boiler has a design capacity of 1200 T/H super-heated steam. To reduce plant electrical house load plant performance team suggested to install VVVF drives on motors for boiler feed water pumps. My responsibility in this task of execution team was to design, develop, test and commission the following systems for new VVVF drives including Graphic User Interface (GUI), Protection system, pump sequence control system, VVVF drive bypass system, speed control system, drum level control via VVVF drives, PID controller tuning without any support from Original Equipment Manufacturer (OEM).

The logic development was planned according to the control philosophy. The control and protections parameters were calculated as there was no data available from pumps OEM for VVVF drives operations. The drum volume, hold time and other controlling parameters were also calculated. Based on the process behavior, first time adaptive control was used for drum level control via VVVF drives.

The results of the control logic was very successful. The total deviation in the frequency control signal was **0.8 Hz** at stable load which is a smooth signal for VVVF drive and pump operation also. At stable load, feed water flow standard deviation remained within 2%. The control of drum level on full load was excellent and drum level was with in **10mm** band and control frequency was within **0.6 Hz** range.

This is the **first high level project** of our plant in which we planned, design, develop and commissioned the system without any support from OEM. This is the **first and only project in Pakistan** in which adaptive control is used for Drum level control system and feed water pumps VVVF drives cut in /cut out automatically while controlling the drum level on frequency. There was not a signal failure in the logic execution and sequence system. The main goal of the project was to reduce the unit auxiliary load and the maximum reduction in unit auxiliary load was about **5 MW** at 50% load and **3 MW** at 100% load.

Table of Contents

Table of Figures

Introduction:

Background:

Lalpir Power Limited is an HFO fired thermal power plant having capacity of 365MW. The boiler has a design capacity of 1200 T/H super-heated steam. There are three boiler feed water pumps each has a design flow capacity of 750 m^3/H @ 220 Kg/cm^2. The unit load can vary from 84 MW to 365 MW as per the National Power Control Center (NPCC) demand. Hence feed water demand also varies from 300 T/H to 1200 T/H according to the steam flow demand. At lower load (300 T/H) only one feed water pump is sufficient for boiler drum level. However, for full load (1200 T/H) two feed water pumps are required to maintain the feed water requirements. The third pump remains available as standby for the system. The all three pumps have individual minimum flow valves and discharge valve, but a common discharge header. All three feed water pumps operate directly through 11KV motors each has a capacity of 6200 KW. The boiler drum level is controlled via level control valve (LCV) installed on common discharge header line. Fig. 1 shows an overview of the system.

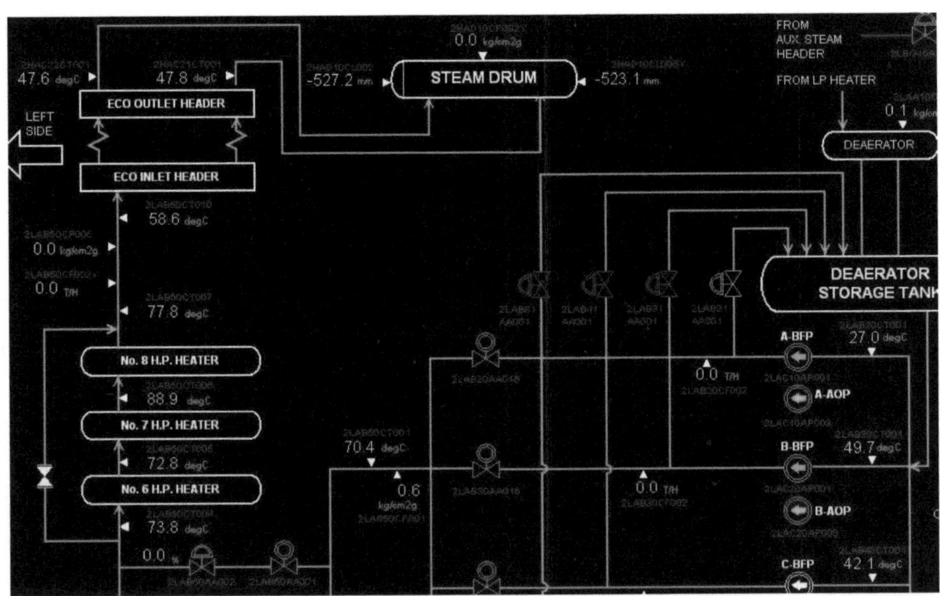

Figure 1. Existing feed water system

To reduce the plant auxiliary load, plant performance team suggested to install Variable Voltage Variable Frequency (VVVF) drives on motors for boiler feed water pumps. The VVVF drives of Schneider Electric model # ATV1200-A2800-6666B5S were selected. Engineering team was given the task to hook up these drives with plant Distributed Control System (DCS) model **Diasys Netmation** of **Mitsubishi Hitachi Power Systems** (MHPS) Japan.

Responsibilities & Challenges:

I was assigned to lead this challenge with team of two I&C Engineers. My main responsibilities were to design, develop, test and commission following systems for new VVVF drives.

- Graphic user interface
- Protection system
- Pump sequence control system
- VVVF drive bypass system
- Speed control system
- Drum level control via VVVF drives
- PID controller tuning

This was first major project in a sense that our engineering team was going to attempt execution in-house without OEM support. We faced following main challenges during the execution of the project:

1. DCS OEM (MHPS JAPAN) refused to share any information & guidance for VVVF hook up with the DCS system.

2. Feed water pumps OEM (WEIR Group) denied to provide any data to operate the pumps on VVVF system.

3. There was no prior reference available in Pakistan to run the feed water system automatically at VVVF system.

4. There was no data available for minimum flow of feed water pumps against pump speed.

5. We had to develop new logic in existing DCS with an option to bypass the new control logic to run the pumps on old control logic as bypass of VVVF drives with a single click at GUI.

6. The feed water pumps VVVF drives should cut in / cut out automatically depending on feed water demand.

7. In existing logic boiler drum level was being controlled by regulating the feed water flow through opening of control valves. In new logic, boiler feed water had to be controlled by changing the boiler feed pumps motor frequency. However level control valve must act as backup if drum level increase sharply due to malfunction of VVVF or load shedding due to electrical grid power supply interruption.

8. If one feed pump trips and standby pump fails to start when load is more than 50%, unit 'Runback' should occur and unit load drop to 180 MW.

9. There should be a system which optimize the boiler feed water pump performance at various speeds when operating in parallel.

10. The standard deviation in feed water flow must be within 2% at stable load to comply with ASME PTC 6 code.

Planning

Planning is an important phase in project management. All the required information was collected from control room and other relative departments to form the control philosophy. Several meeting were conducted with control room engineers (CRE) to finalize the graphic user interface schemes. There were following two main planning categories.

Hardware Planning:

A list of inputs / outputs was prepared for VVVF drives and hardware being reserved in the DCS. Procurement process was also initialized for the purchase of required hardware. The I/O list of one boiler feed water pump VVVF drive is as under.

DESCRIPTION	I/O TYPE	I/O #	SYSTEM	CUB #	TBU #
VFD SPEED FEED BACK	AI	AI-1007	APC	CUB#3	3TBUR4-TB1-13,14
VFD OUTPUT CURRENT	AI	AI-1008	APC	CUB#3	3TBUR4-TB1-15,16
VFD SPEED DEMAND	AO	AO-1023	APC	CUB #3	3TBUR8-TB1-13,14
VFD SPEED DEMAND	AO	AO-1031	APC	CUB #3	3TBUR8-TB1-13,14
VFD MAJOR FAULT	DI	DI-00488	SEQ-1	CUB #4	4TBUR8-TB1-15,16
VFD RUNNING	DI	DI-00489	SEQ-1	CUB #4	4TBUR8-TB1-17,18
VFD READY	DI	DI-00490	SEQ-1	CUB #4	4TBUR8-TB1-19,20
VFD MINOR FAULT	DI	DI-00491	SEQ-1	CUB #4	4TBUR8-TB1-21,22
VFD REMOTE	DI	DI-00492	SEQ-1	CUB #4	4TBUR8-TB1-23,24
VFD BYPASS	DI	DI-00493	SEQ-1	CUB #4	4TBUR8-TB1-25,26
VFD START COMD	DO	DO-00286	SEQ-1	CUB #3	3TBUF7-TB2-27,28
VFD STOP COMD	DO	DO-00287	SEQ-1	CUB #3	3TBUF7-TB2-29,30

Table 1. List of IOs for VVVF Drive Interface with DCS

Software Planning:

The *Distributed Control System* (DCS) in Lalpir power is of MHPS Diasys Netmation. In case of control net (communication media of subsystems) failure, all subsystems are capable of running independently. The DCS consist of following six subsystems which are interconnected with redundant control net.

1. Sequence 1 (Drive protection system for Boiler)

2. Sequence 2 (Drive protection system for Turbine)

3. BMS (Burner Management System)

4. APC (Automatic Plant Control, All Close Loops)

5. DEH (Digital Electro Hydraulic, Turbine Governing System)

6. IPU (In Put Unit, All Open Loops)

The BFP VVVF drives protection system, sequence of operation and bypass system were planned in Sequence 1 subsystem. The APC has VVVF speed control system and drum level control system. There were also some signals that are interconnected in both subsystems. The interconnecting signals were reviewed for *capability of withstanding* in case of failure of control net so that equipment and plant safety can be ensured.

The logic sheet locations were marked for new development and for modifications in existing sheets. DCS database backup was taken before the logic development initialization.

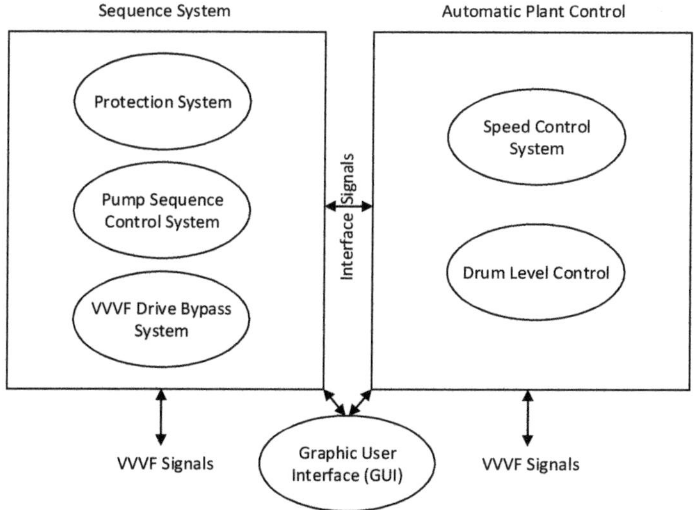

Figure 2. Logic blocks in DCS subsystems

Execution:

Sequential Logic Development

Mode Selection:

First of all, new logic development started form the selection of the RUN mode of BFP. There are two modes of BFP VVVF drives.

1- Bypass mode: BFP will run as per old logic directly from 11KV breaker and the VVVF new logic will be bypassed.

2- VVVF mode: BFP will run through VVVF drive at variable speeds as per process demand. Mode selection logic development is as under.

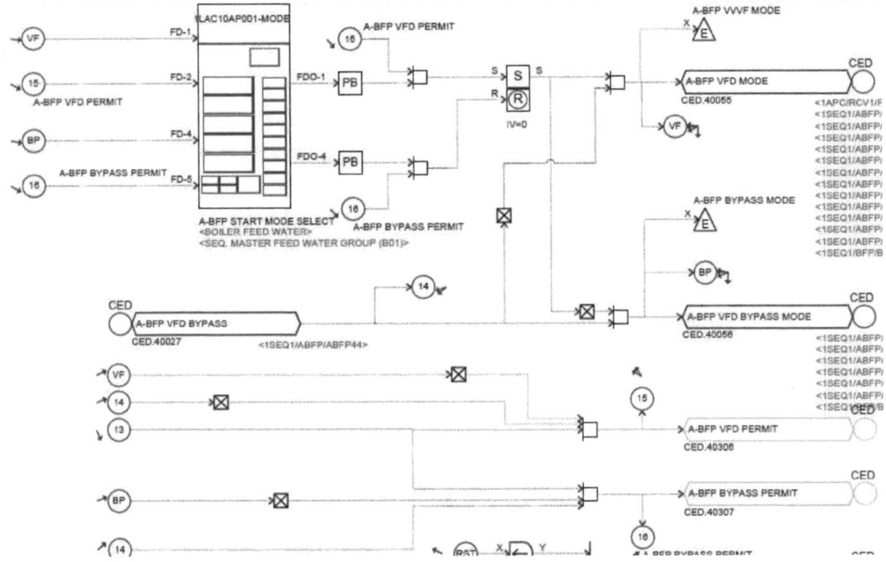

Figure 3. Mode selection logic

There are some protections for mode selection logic to prevent the equipment damage. The mode selection is only possible when 11 KV breaker is in off position. Secondly, if VVVF drive switched to bypass locally, then in DCS VVVF mode selection will turn off even if already at VVVF mode and also will be disabled in DCS.

Bypass mode selection is only possible when VVVF is selected as bypass from local panel and 11KV breaker is in off condition. The graphic user interface for mode selection is shown below.

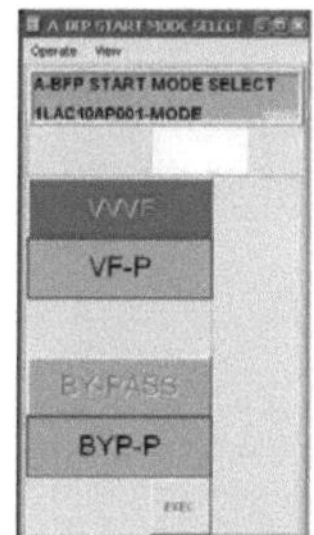

Figure 4. VVVF mode selection loop plate

VVVF Drive ON/OFF Commands:

To turn on VVVF drive, its 11KV breaker should ON first. The logic of ON/OFF command of 11 KV breaker of VVVF is shown below. The 11 KV beaker of each drive has its own separate logic. This logic is linked with 11KV graphic sheet and have a loop plate user interface for control room engineers.

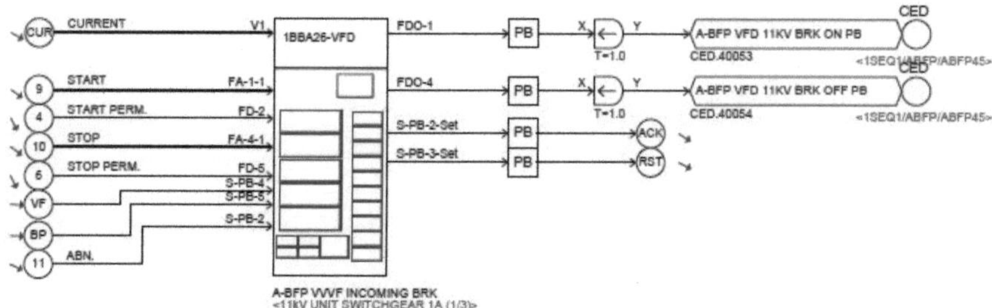

Figure 5. Logic of ON/OFF command of 11 KV breaker

The ON permissive of 11 KV breaker is linked with the VVVF mode selection. The permit will only be available if BFP is selected for VVVF operation otherwise 11KV breaker will be linked with bypass ON command.

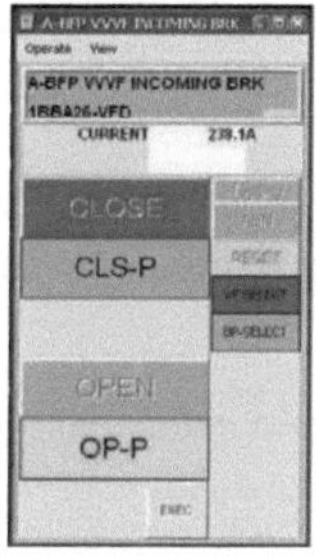

Figure 6. Graphic user interface of 11 KV breaker

The ON permissive of VVVF Drive will be available only if 11KV breaker's ON status is available. The manual operation of VVVF drive is accessible through a loop plate user interface. The logic of manual operation has all the required prerequisites as start permit for the protection of boiler feed water pumps.

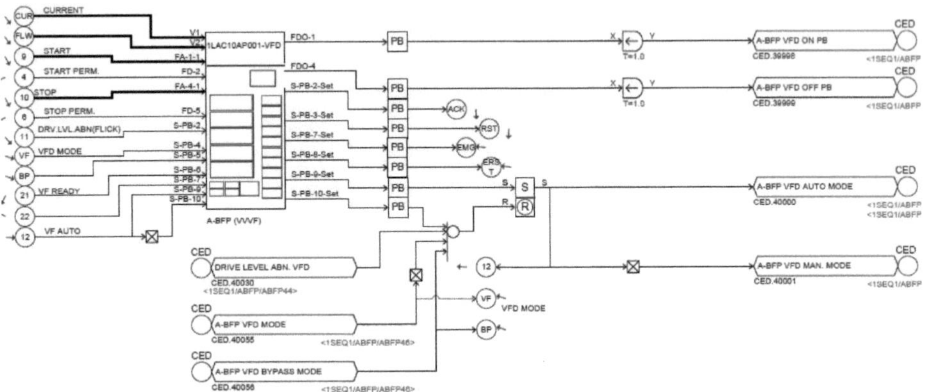

Figure 7. VVVF drive manual ON/OFF command logic

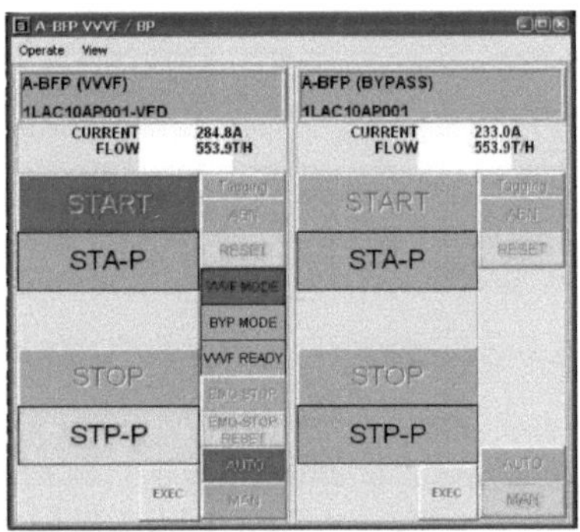

Figure 8. Graphic user interface of VVVF drive ON/OFF command

BFP Minimum Flow Protection:

Operation of centrifugal pumps below their minimum flow requirements is the primary cause of premature pump failure. Hydraulic instability occurs at low flows, causing cavitation, surging, and excessive vibration in the pump.

There was no information available for the required minimum flow at various speeds from the OEM. We only had data of the flow at full speed from minimum flow line which was 247 T/H. The minimum flow protection was on 171 T/H at full speed (50Hz).

Following formulas were used to calculate the flow curve and pressure with respect to frequency / speed of the pump. By using MS excel, graphs were drawn to set the minimum flow line at 05 T/H less than actual. The line head pressure and Dearater pressure were also compensated for minimum flow line that is connected to the Dearater tank.

Flow is proportional to shaft speed:

$$\frac{Q_1}{Q_2} = \left(\frac{N_1}{N_2}\right)$$

Pressure or Head is proportional to the square of shaft speed:

$$\frac{H_1}{H_2} = \left(\frac{N_1}{N_2}\right)^2$$

Where

Q is volumetric flow rate

N is shaft rotational speed

H is pressure or head developed by the pump

Speed	Pressure	Flow	Trip
0	0	0	0
2.5	0.55	12.35	0
5	2.2	24.7	0
7.5	4.95	37.05	0
10	8.8	49.4	44.4
12.5	13.75	61.75	56.75
15	19.8	74.1	69.1
17.5	26.95	86.45	81.45
20	35.2	98.8	93.8
22.5	44.55	111.15	106.15
25	55	123.5	118.5
27.5	66.55	135.85	130.85
30	79.2	148.2	143.2
32.5	92.95	160.55	155.55
35	107.8	172.9	167.9
37.5	123.75	185.25	171
40	140.8	197.6	171
42.5	158.95	209.95	171
45	178.2	222.3	171
47.5	198.55	234.65	171
50	220	247	171

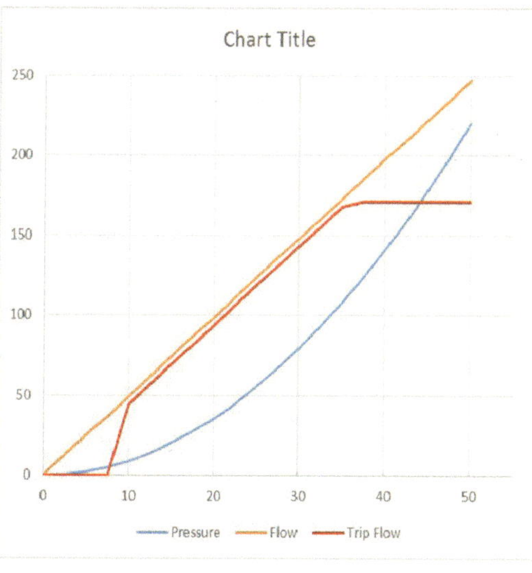

Figure 9. Minimum Flow Protection Calculation & Graph

The implementation of the calculation was accomplished by modifying the existing logic. If the VVVF mode was selected then minimum flow will be calculated with respect to the speed. Otherwise it will be fixed at 171 T/H. The pump will trip if feed water flow drops below minimum flow for 5 seconds. The logic development is shown below.

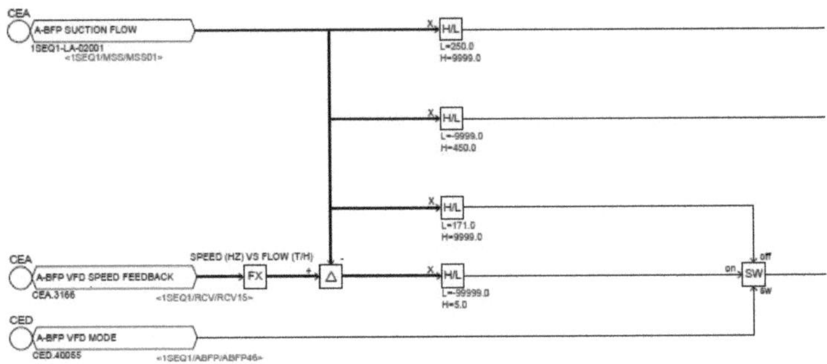

Figure 10. Minimum Flow Protection on VVVF mode

Sequence System:

The master sequence of all three BFP VVVF drives follows as:

If sequence is ON, the BFP VVVF drives will cut in automatically as per process demand or in case of tripping of any BFP VVVF drive. Control Room Engineers can also change the duty of the BFP by selecting the priority of the pump from sequence user interface. For duty change over sequence, new selected pump will start and will reach the 34Hz. After that it will follow the process demand and on duty pump will decelerate to 35 Hz and turn off. The changeover will be smooth without any disturbance to the drum level.

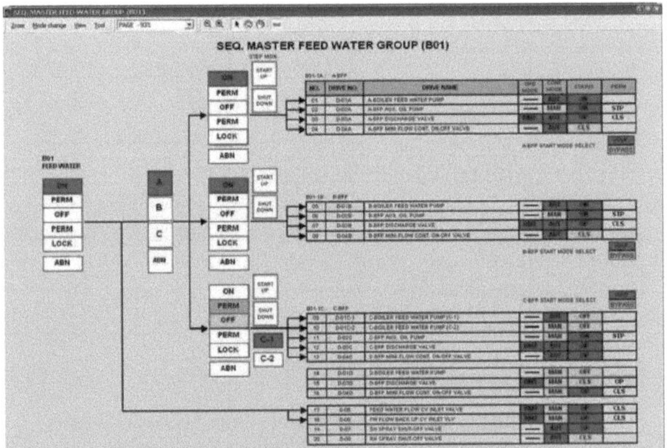

Figure 11. Sequence user Interface

For smooth changeover of BFP VVVF drive duty, the logic was developed in such a manner that changeover initiate at a certain point of 34Hz. This logic of changeover point is as under and named as ON END STATE. The changeover will begin after ON END STATE which is 34 Hz speed.

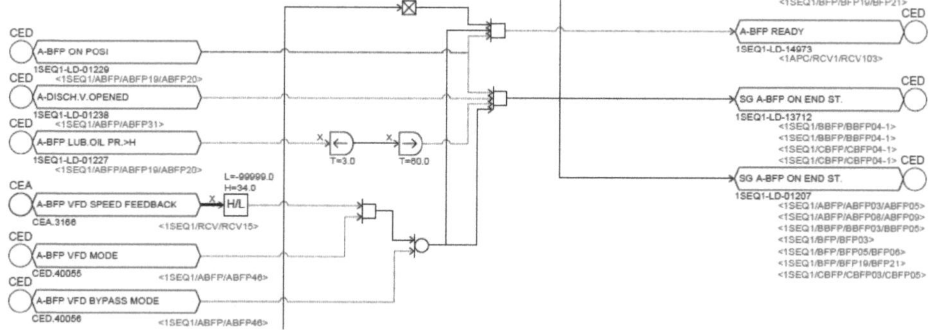

Figure 12. VVVF ON END STATE

After completing the logic development in SEQ system, the graphic elements were linked the to the user interface of main feed water system. The BFP VVVF drives main user interface changed as shown below.

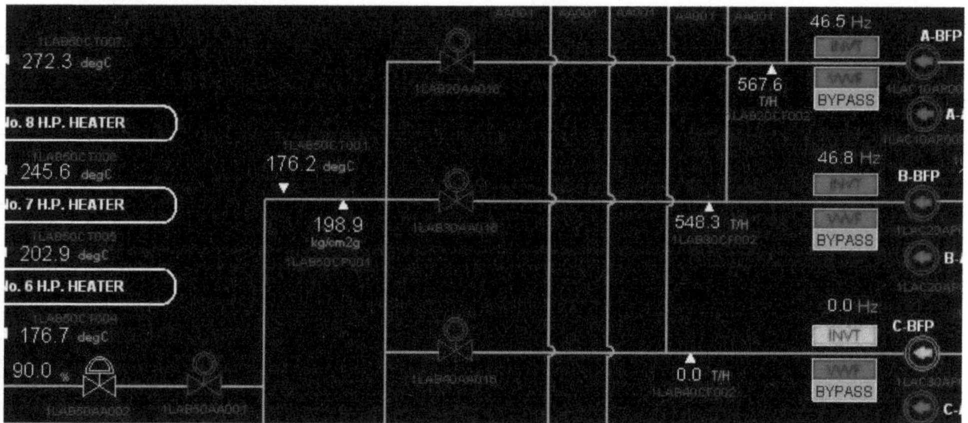

Figure 13. Feed water system main user interface

There were more than sixty new logic sheets which were developed. However, only main portions of the logic are shown in this report.

Control Logic Development

Below is the procedure elaborated that was used to develop the boiler drum level control system:

Boiler Drum Volume:

The first thing that had to be calculated for drum level control system was the drum volume in control range. It had a vital role for the calculation of hold up time of boiler drum at various conditions. To calculate the drum volume, a simple technique was used to fill the drum with constant flow rate then note down the time difference of -300mm level and 0mm level.

As shown in trend, it took 3.03 minutes to reach the boiler level from -300mm to 0mm at 1.9 T/min flow rate. So the drum volume for 300mm will be 1.9 x 3.03 = 5.76 Tons. The low level protection is operated at about -400mm level. So the trip volume will be 7 Tons approximately as proportion to above calculations.

Drum Volume Calculations		
	T/Hour	T/Min
Flow Rate	114	1.9
time for -300mm to 0.0 mm	3.03	Minutes
Drum Volume (control range)	5.757	Tons
Drum Volume (trip range)	7	Tons

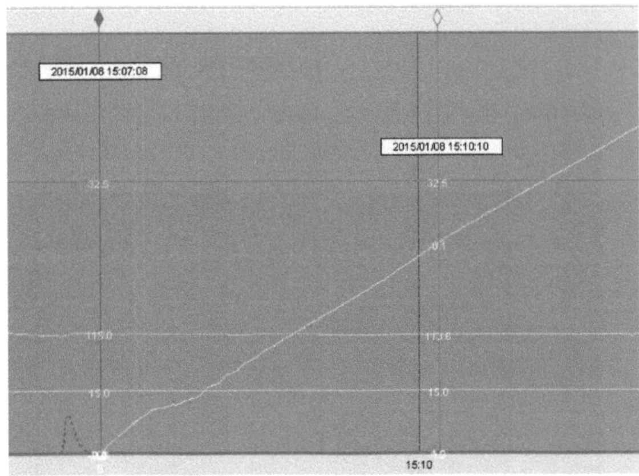

Figure 14. Drum Volume Calculations

Boiler Drum Hold up Time:

The boiler drum hold up time is very important factor for control system design. It depends upon steam conversion rate and drum volume. As the drum volume is fixed for all the unit loads so the drum hold up time will decrease as steam flow increase if the intake feed water flow is contant. However, the intake feed water flow also varies as second pump cut in after 50% unit load.

Now there are two scenarios, if the single pump trips after 50% load when two pumps are in service or the only runing pump trip below the 50% load. Both were calculated in following table and shown graphically.

Main Steam Flow T/H	Drum Hold Up time (both pump trips) Seconds	Drum Hold Up time (one pump trips) Seconds
0		
50	504	
100	252	
200	126	
300	84	
400	63	
500	50	
620	41	
700	36	1260
750	34	360
800	32	210
900	28	115
1,000	25	79
1,100	23	60
1,200	21	48

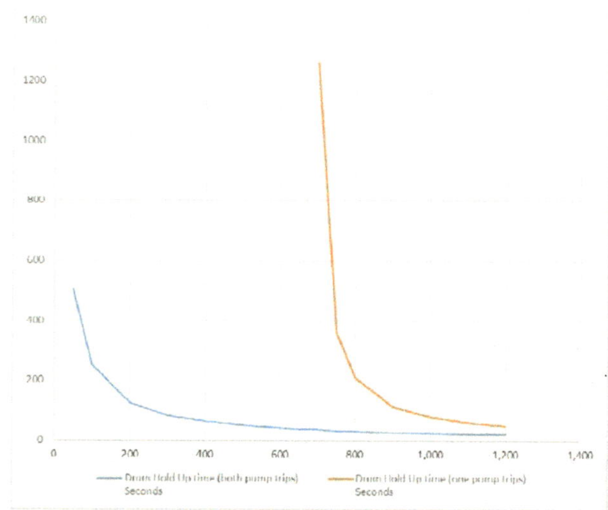

Figure 15. Boiler Drum Hold up Time

At a specific load, If system delays to add feed water in the boiler drum more than hold up time, then there could be a huge damage to boiler tubes and other structure. To avoid this situation, system should reduce the unit load to decrease the steam generation rate so that boiler drum hold up time could be increased. This protection is known as a unit run back protection.

Unit Run Back Logic:

If the unit load is more than 180 MW and one pump trips and the standby pump unable to start in 5 seconds then unit will shed the load @ 1086 MW/Min and will stay at 180 MW. Although in VVVF scheme the pump starts in 4 seconds however, pump will not contribute 100% to the feed water system for about 57 seconds as per the timing test results shown below. This is because of ramp rate of the VVVF drives from OEM.

The drum hold up time is about 48 seconds at full load. Whereas VVVF will take about 57 seconds so the boiler will be at high risk as drum will experience very low level which could cause boiler tube damage due to excessive overheating. To avoid the high risk it is decided that unit will run back if any VVVF trip on full load for the safety of the boiler. This is a weak point of VVVF drive system that it is unable to support the process at full load in short time. Hence it is the system limitation.

VVVF Drive Timing Test

Test	Description	Time (Seconds)
ON Command	Running feedback after start command	4
	ON Abnormal Setting	7
Acceleration	Start Command to 35 Hz	37
	35 Hz to 50 Hz	20
Deceleration	50 Hz to 35 Hz	25
	OFF Command at 50 Hz to OFF Indication	26
	OFF Abnormal Setting	30

Boiler Drum Water Consumption Rate:

To design the control logic for any process, it is essential to understand the process behavior. The primary object of the control logic is to keep up the drum water level at its set point. System can maintain the drum level only if it is able to maintain feed water according to consumption. Due to sudden process disturbances and increasing process gain when second pump cuts in it is very difficult to tightly control the required drum level so it is necessary to alter the control loop parameters according to the process behavior.

To analyze the consumption rate, steam conversion rate was calculated inside the boiler drum as shown in following table & graph.

Steam Flow	Consumption Rate Tons / Second
300	0.0833
400	0.1111
500	0.1389
600	0.1667
700	0.1944
800	0.2222
900	0.2500
1000	0.2778
1100	0.3056
1200	0.3333

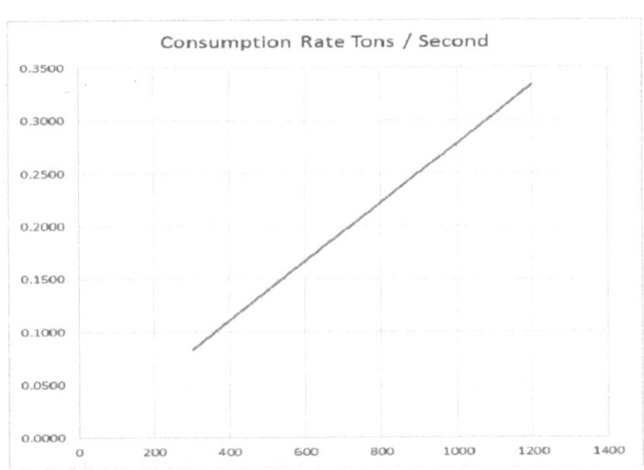

Figure 16. Boiler Drum water Comsumption

The graph indicates that consumption rate at minimum load (300 T/H) is 0.0833 T/s while at full load (1200 T/H) it increase up to 0.3333T/s. This increases consumption rate up to 4 times as the main steam flow increases to full load demand. So the total gain of the control loop should be increased accordingly to mitigate the increasing demand of the drum level for a tight drum level control.

Drum Level Controller Calculations:

The process parameters for minimum and maximum load are as under.

Description	Minimum load Value	Maximum load Value
Drum Pressure (kg/cm^2)	165	188
VVVF Min Freq. to produce the drum pressure(Hz)	43.5	46
drum control volume (Tons)	7	7
drum water consumption rate (Tons/seconds)	0.083	0.333
Main steam flow (T/H)	300	1075
Boiler drum hold up time (Seconds)	84	21

As it is obvious from the table that there is a huge change in process parameters from minimum to maximum load. The drum water consumption rate increased up to 4 times so it is crucial to make up the drum water at same pace for tight control which was very difficult by using fixed gains for control loops.

To design and calculate the adaptive control as per process demand, following books were consulted for effective implementation of the system.

Reference Books:

1. **BASIC AND ADVANCED REGULATORY CONTROL: SYSTEM DESIGN AND APPLICATION By Harold L. Wade**

2. **PROCESS CONTROL: A PRACTICAL APPROACH By MYKE KING**

First of all, feed water controller tuning was performed at minimum load by using Ziegler-Nichols close loop method. The output of the level controller and main steam flow signal was forced to maintain a fixed set point for feed water controller. Then feed water controller gain was increased up to a level where a sustainable oscillation in feed water flow observed.

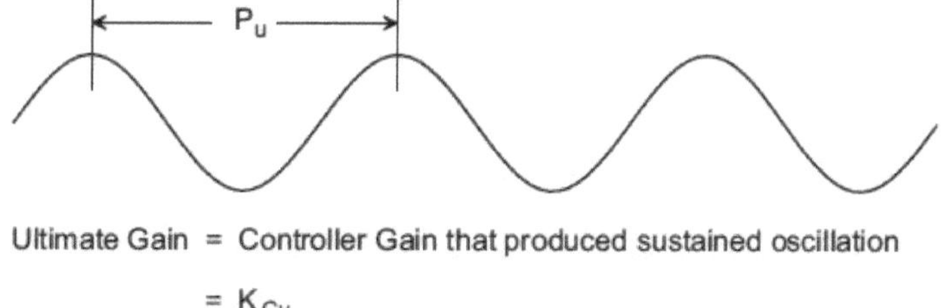

Ultimate Gain = Controller Gain that produced sustained oscillation

= K_{Cu}

Ultimate Period = P_u

Figure 17. Ultimate gain oscilation

The gain and integral time were calculated by using following formulas. The same procedure was applied on level controller by fixing the main steam flow and feed water flow signals.

Proportional + Integral (PI)	Gain	$K_C = 0.45\ K_{Cu}$
	Integral Time (minutes/repeat)	$T_I = 0.83\ Pu$

The calculated results were not 100% accurate. The controller parameters were fine tuned to eliminate the minor variations in process parameters. There was minor variations in the main steam flow signal as it is calculated indirectly by using turbine first stage discharge pressure. To eliminate this effect, level controller gain increased and feed water controller gain decreased. However, total loop gain was remain constant.

The level controller gain decreased as load increase while feed water controller gain decrease initially to maintain the process parameters when second pump cut in. However, as feed water controller should act before actual change in drum level, its gain increase after pump cut in to 3 times to cope up the increased demand of feed water in drum level efficiently.

The total loop gain increased up to 1.5 times. However, it's resulted as 3 times increased as two pumps used to be in service parallel at full load. Further details of adaptive control are in next chapter.

Boiler Drum Level Adaptive Control:

At our plant, OEM used fixed numbers for DCS controller's gain. However, it is very difficult to control the drum level on every load point by using the fixed gain for loops because process parameters for minum load and maximum load are different. As indicated earlier that feed water pumps will cut in as per load demandthe cut in of feed water pump will disturb the feed water flow significantly hence could causing the uncontrolled cycling of feed water flow and drum level. Based on these process conditions, it was decieded to use adaptive control so that it could change the control loop gain according to the process requirement.

The gain of drum level and feed water loop was linked with main steam flow. As the main steam flow increases the gain of the controller also chages according to the defined parameters as defined below.

Total Gain = Drum Level Gain * Flow Loop Gain

Main Steam Flow	Drum Loop Gain	Flow Loop Gain	Total Gain
0	1.600	0.0100	0.0160
50	1.600	0.0100	0.0160
100	1.600	0.0100	0.0160
200	1.600	0.0100	0.0160
300	1.600	0.0100	0.0160
400	1.600	0.0100	0.0160
500	1.300	0.0100	0.0130
620	1.240	0.0085	0.0105
700	1.200	0.0085	0.0102
750	1.180	0.0085	0.0100
800	1.160	0.0090	0.0104
900	1.120	0.0100	0.0112
1,000	1.080	0.0125	0.0135
1,100	1.040	0.0230	0.0239
1,200	1.000	0.0300	0.0300

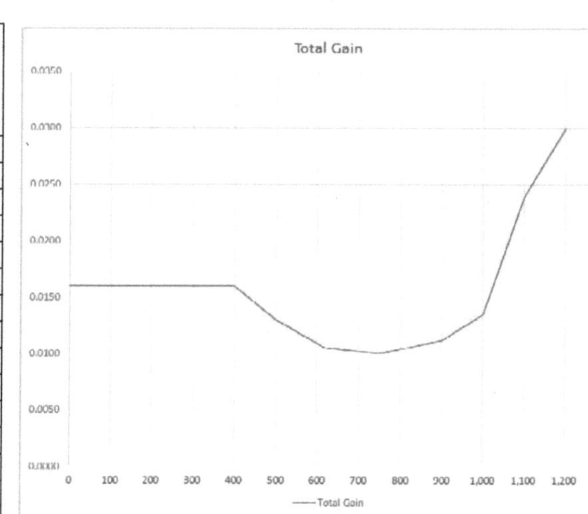

Figure 18. Boiler Drum Level Loop Gain

At lower loads, the gain of level loop is higher and decreases gradually as load increases. However, the feed water flow controller gain decrease initially to accommodate the variations causing by the pump that cut in at 620 T/H steam flow. After the pump cut in the gain of feed water flow controller increases to meet the process demand.

At higher loads the feed water flow controller take the control as feed farward loop so that the change in drum level can be anticiptated before the actual disturbance. The logic below shows the pratical implementation of the adaptive conrtol for drum level.

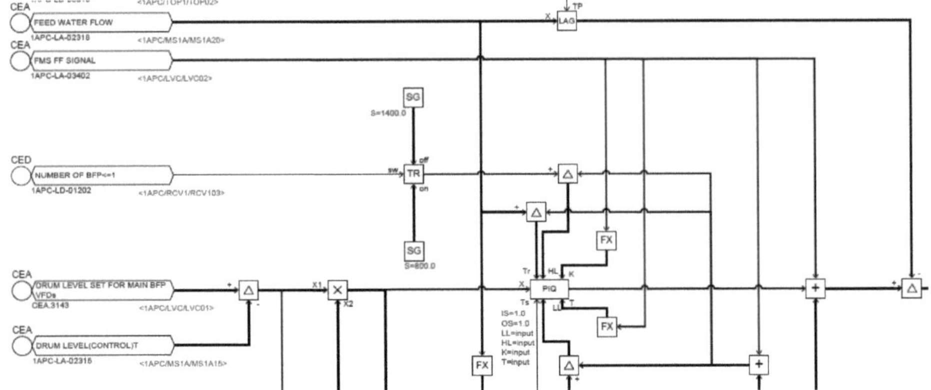

Figure 19. Drum Level Controller

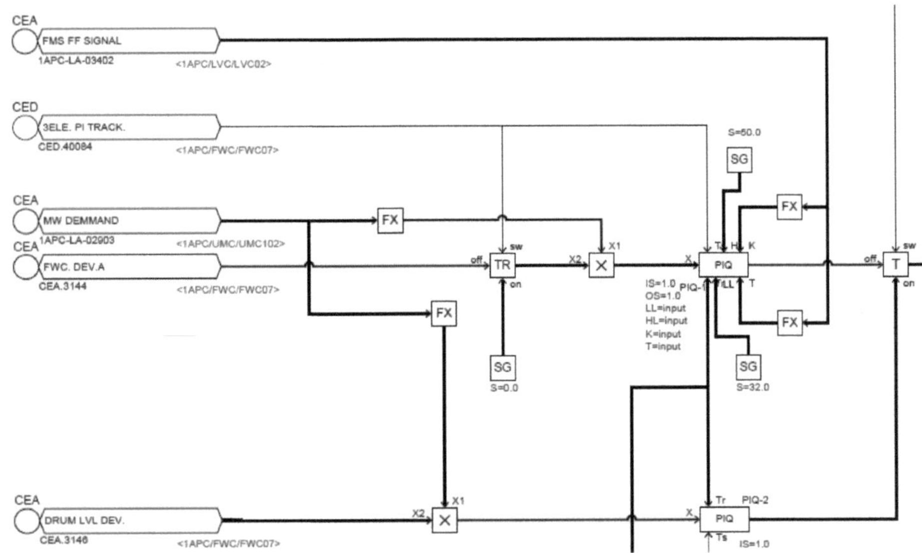

Figure 20. Feed water Controller

Standby Drum Level Control:

To achieve the maximum power saving from VVVF drive the drum level control valve must fully open to decrease the feed water flow resistance. However, in case of VVVF drive malfunction the level control valve should act as final control element to prevent the drum level overshoots. The process behavior was analyzed and the logic was developed so that the control valve remain open to 90% if _all_ following conditions fulfill. Otherwise it will act as per 3 element control of drum level.

1. Any VVVF drive is on Auto mode.

2. Any VVVF drive is ON.

3. Main Level Control Valve selected for control and on AUTO

4. The drum Level remain in control below +150mm level.

If the control room engineer puts all VVVF drives on manual, the level control valve will immediately take control to prevent the deviation in drum level. The practical implementation of the above logic is as under:

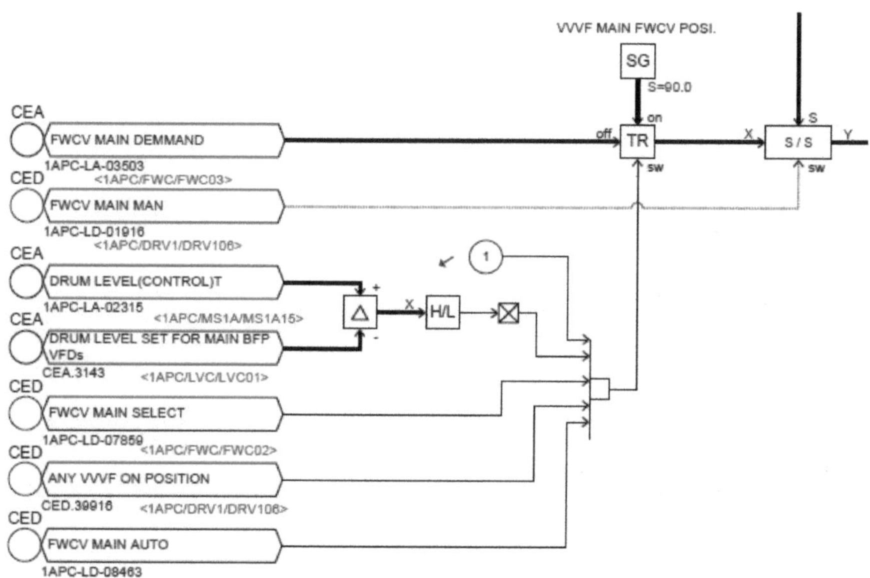

Figure 21. Main Level Control Valve Control

DCS Self-sustainability:

As mentioned earlier, there were interconnected signals between APC and SEQ systems of DCS. The DCS sub systems have an ability to work independently in case of control net communication failure. . The interconnected signals were analyzed to ensure DCS self-sustainability.

In APC the VVVF Drive ON signal is coming from SEQ system. This signal is used to pass the frequency demand to VVVF drive. If this signal fails the frequency demand to VVVF will be zero. The same scenario will develop if the communication between DCS subsystems fails and as a result all VVVF drive will drop to minimum frequency causing the low feed water flow and low drum level and hence damage to the boiler tubes.

A flip flop is used in APC to latch the SEQ1 VVVF Drive ON signal in case of communication failure. The ON signal will be held by latch unless an off signal from SEQ system is received. If all signal from SEQ system fails while the drive is ON all signals including ON & OFF signal will fail and APC will hold the last ON signal to prevent the VVVF drive from getting OFF.

The logic of APC system is shown below for reference.

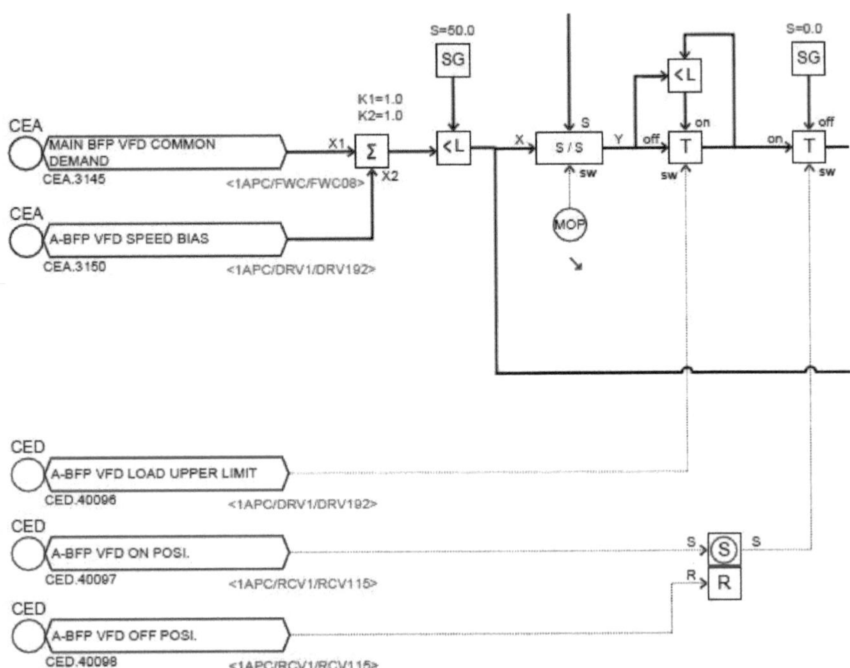

Figure 22. APC VVVF Drive ON Signal

Simulation & Commissioning:

After the completion of logic development, simulation testing procedure was planned and prepared as shown below:

Field testing was completed of all I/Os including cable resistance and contact resistance. Then field simulation of I/O was carried out. The logic was then loaded to DCS. The DCS has an option to simulate the loops while keeping the 11 KV breakers in test position.

All user interfaces and loop plates including sequences were tested as per the planned documents and signed by the control room engineer for its functionality & conformity.

Boiler feed water pumps were tested on minimum flow line successfully. The calculated minimum flow trip logic was also tested successfully. All 11 KV breakers and VVVF drives protections were tested. VVVF & bypass modes were also tested independently.

1	**Selection mode VVVF**	i) BFP-A Select on VVVF Mode from Loop plate	BFP-A VVVF will Select
		ii) BFP-B Select on VVVF Mode from Loop plate	BFP-B VVVF will Select
		iii) BFP-B Select on VVVF Mode from Loop plate	BFP-C VVVF will Select
2	**i) BFP-A 11KV breaker ON/OFF**	i) BFP-A 11KV ON command from Loop plate given in 11KV Graphics	i) BFP-A 11KV breaker will ON
			ii) BFP-A ON permit will
			iii) BFP-A OFF permit will
			iv)BFP-A VVVF transformer will
		ii) BFP-A 11KV OFF command from Loop plate given in 11KV Graphics	i) BFP-A 11KV breaker will OFF
			ii) BFP-A ON permit will appear
			iii) BFP-A OFF permit will
			iv)BFP-A VVVF transformer will
3	**ii) BFP-B 11KV breaker ON/OFF**	i) BFP-B 11KV ON command from Loop plate given in 11KV Graphics	i) BFP-B 11KV breaker will ON
			ii) BFP-B ON permit will
			iii) BFP-B OFF permit will
			iv)BFP-B VVVF transformer will
		ii) BFP-B 11KV OFF command from Loop plate given in 11KV Graphics	i) BFP-B 11KV breaker will OFF
			ii) BFP-B ON permit will appear
			iii) BFP-B OFF permit will
			iv)BFP-B VVVF transformer will

Table 2. Sample of simulation & testing procedure

Results & Achievements:

The system started very smooth. The flow curve of the pumps was according to the calculated one. The minimum flow curve was also matched with anticipated one. The control of the drum level was excellent. At stable loads, the total deviation in the frequency control signal was **0.8 Hz** which is a smooth signal for VVVF drive and feed water pump operation. The feed water flow standard deviation remained within 2%. The control of drum level on load ramping and on full load was excellent and drum level was with in **10mm** band and control frequency was within **0.6 Hz** range at full load.

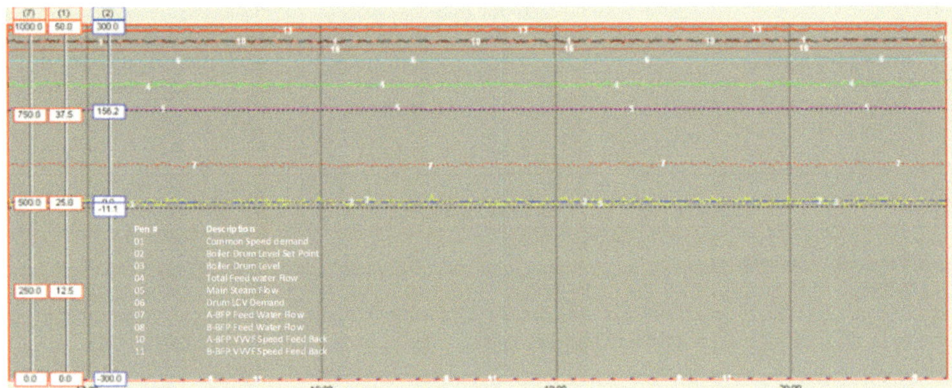

Figure 23. VVVF drive drum level control at full load

At 50% load where second pump cuts in, the total deviation in drum level and feed water flow was well within the limit. The drum level deviation remained in **20mm** band where as in old logic its deviation was about **50mm**. At same frequency, there was a difference in performance of the pumps. So performance mismatched issue resolved by using the frequency bias system.

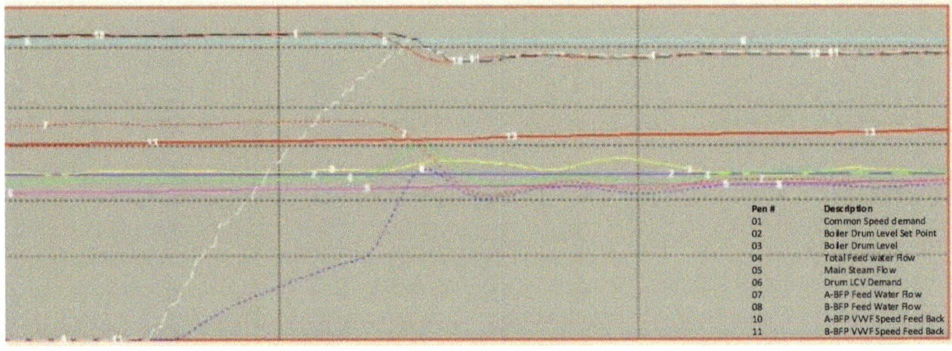

Figure 24. Auto Cut In of BFP

The most critical part was the duty change over sequence at full load. In duty change over sequence, stand by pump has to start and one in service pump will be stopped as per priority sequence. The changeover of pumps was smooth with no deviation in drum level. Control loop showed excellent control while the changeover of the pumps.

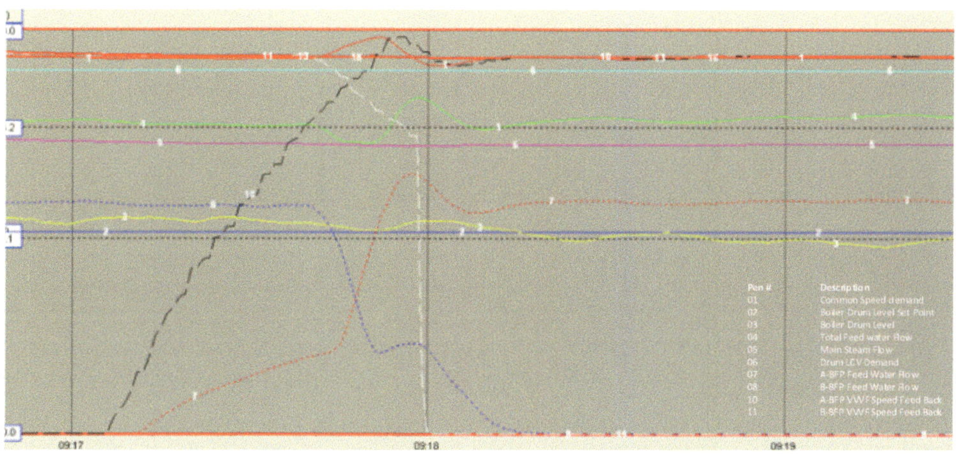

Figure 25. VVVF drive duty change over sequence

The drum level control of old logic developed by OEM (MHPS Japan) was on fixed gain and has an average control on full load. Whereas by using the adaptive control in the new control logic, drum level control become tight and smooth. The difference of the control is obvious in following trends.

Initially the drum level control was on Level control valve (OEM logic). Then it transferred to the new logic.

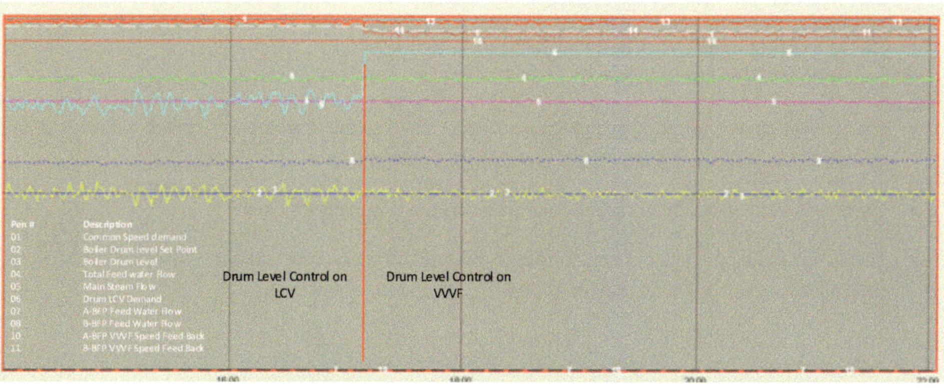

Figure 26. Old Control Logic *VS* New Control Logic

The auto cut out of VVVF drive while decreasing of the load was also very successful. There was no deviation observed in drum level where as in feed water flow only minor cycling observed.

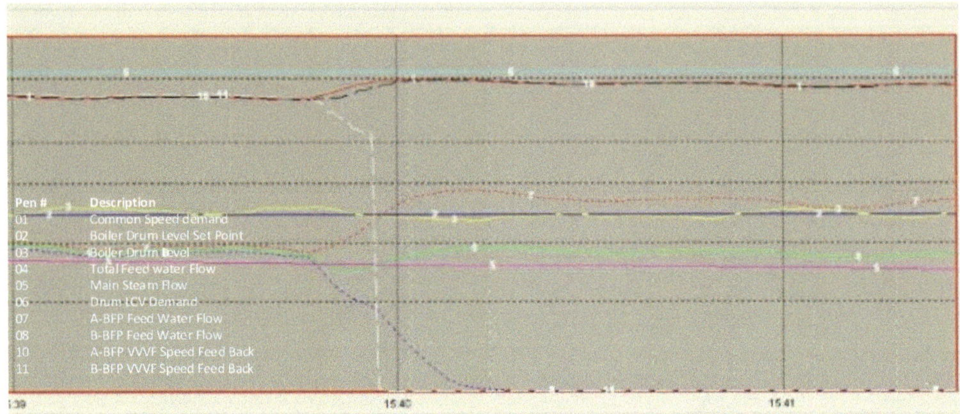

Figure 27. BFP VVVF Drive Auto Cut off

Over all the project was highly successful. The highlights of the project are as under.

- This is the **first high level project** of our plant in which we planned, designed, developed and commissioned the system without any OEM support.

- This is the **first and only project in Pakistan** in which adaptive control is used for such capacity boiler drum level control system.

- This is **first & only project in Pakistan** in which feed water pump VVVF drives cut in /cut out automatically while controlling the drum level on frequency.

- There was not a signal failure in the logic execution and sequence system.

- The main goal of the project was to reduce the unit auxiliary load and the maximum reduction in unit auxiliary load was about **5 MW** at 50% load and **3MW** at full load.